MENTAL MATH FOR
MATH HATERS

Table of Contents

Introduction 3
Chapter 1: Understanding Our Number System 9
Chapter 2: Addition Strategies 14
Chapter 3: Subtraction Strategies 20
Chapter 4: Multiplication Basics 27
Chapter 5: Advanced Multiplication Techniques 34
Chapter 6: Division Shortcuts 41
Chapter 7: Percentages and Fractions 47
Chapter 8: Estimation and Approximation 56
Chapter 9: Mental Math in Daily Life 63
Chapter 10: Memory Techniques for Math Facts 71
Chapter 11: Overcoming Math Anxiety 80
Chapter 12: Fun with Numbers 87
Conclusion: Your Ongoing Math Journey 93
Glossary of Terms 99

Introduction

My Journey from Math Phobia to Math Enthusiasm

As I sit down to write this book, I can't help but chuckle at the irony. Me, writing a math book? If you had told my 12-year-old self that this would happen, I would have laughed in your face – and then probably burst into tears.

You see, math and I weren't always on speaking terms. In fact, for most of my early life, math was my arch-nemesis, the supervillain in the story of my education. I still remember the cold sweat that would break out on my forehead when a teacher called on me to solve a problem at the board. The numbers would dance before my eyes, mocking me, as my mind went completely blank.

It wasn't just in the classroom. Math anxiety followed me everywhere. Calculating a tip? Panic. Figuring out a discount during a sale? Sheer terror. I even avoided certain career paths because they involved "too much math." I had convinced myself that I was simply "bad at math," and that was that.

But here's the thing about stories – they can always take an unexpected turn.

My transformation began in my late twenties, and it started with something as simple as a puzzle book. I had picked it up on a whim during a long flight, and to my surprise, I found myself enjoying the number puzzles. They weren't like the math problems I remembered from school. These were playful, engaging, and – dare I say it? – fun.

Intrigued by this new feeling, I started to explore more. I discovered YouTube videos that explained math concepts in ways I had never considered before. I found books that approached numbers not as intimidating abstractions, but as useful tools for understanding the world around us. Slowly but surely, I began to see math differently.

The turning point came when I realized that mental math wasn't about being a human calculator. It was about understanding numbers, seeing patterns, and finding clever shortcuts. It wasn't about speed or perfection – it was about flexibility and creativity. This was the kind of math I could get behind!

As I practiced these mental math techniques, something miraculous happened. The world around me began to change. Suddenly, I could split a dinner bill without reaching for my phone. I could estimate my grocery total as I shopped. I could help my niece with her homework without breaking into a cold sweat.

But the most significant change was in how I saw myself. I was no longer "bad at math." I was someone who could learn, improve, and even enjoy working with numbers. This shift in self-perception spilled over into other areas of my life, boosting my confidence and opening doors I had previously thought were closed to me.

That's why I'm writing this book. Not because I'm a math genius (I'm far from it!), but because I've been where you are. I know what it's like to feel defeated by numbers, to think that mental math is a skill reserved for a select few. And I'm here to tell you that it's not true.

This book is the resource I wish I had all those years ago. It's designed for anyone who's ever felt math anxiety, for adults returning to numbers after years away, for parents who want to help their children, and for anyone who simply wants to improve their mental math skills.

On these pages, you won't find dry formulas or intimidating equations. Instead, you'll discover friendly strategies, real-life applications, and a gentle, step-by-step approach to building your confidence with numbers. We'll take this journey together, celebrating small victories and building up to bigger challenges.

Remember, every math expert was once a beginner. Your past experiences with math don't define your future relationship with it. It's never too late to rewrite your math story. So, are you ready to turn the page and start a new chapter?

Let's embark on this adventure together – from math fear to math cheer!

Embracing a New Math Mindset

Welcome to a new way of thinking about math! If you've picked up this book, chances are you're looking to improve your mental math skills, perhaps with a mix of curiosity and apprehension. Let me assure you: you've come to the right place. This introduction will set the stage for your journey, addressing some common concerns and highlighting the exciting path ahead.

The Elephant in the Room: Math Anxiety

First, let's talk about something that affects millions of people worldwide: math anxiety. If you've ever felt your palms sweat or your heart race at the thought of doing math, you're not alone. Studies suggest that up to 93% of adult Americans experience some level of math anxiety. It's so common that it's almost considered normal.

But here's the truth: math anxiety isn't innate. It's learned, often through negative experiences or societal messages. And if it's learned, it can be unlearned. Throughout this book, we'll not only build your math skills but also work on dismantling the anxiety that may have held you back.

Debunking the "Math Person" Myth

How many times have you heard someone say, "I'm just not a math person"? Maybe you've said it yourself. It's time to put this myth to rest. There's no such thing as a "math person" or a "non-math person." Our brains are all capable of learning and improving mathematical skills.

Recent research in neuroscience shows that our brains exhibit plasticity - the ability to form new neural connections throughout life. This means that with practice and the right approach, anyone can get better at math, regardless of age or past experiences.

The Power of Mental Math in Daily Life

You might be wondering, "Why bother with mental math in the age of smartphones?" Great question! While calculators are undoubtedly useful, relying solely on them can be limiting. Mental math offers several unique benefits:

1. Practical Convenience: Calculate tips, discounts, or split bills quickly without reaching for your phone.
2. Improved Estimation Skills: Develop a better sense of numbers, helping you catch errors and make quick decisions.
3. Enhanced Problem-Solving: Mental math improves overall logical thinking and problem-solving abilities.
4. Boosted Confidence: As your skills improve, you'll feel more confident in various life situations.
5. Brain Exercise: Mental math serves as a great workout for your brain, potentially helping to keep it sharp as you age.

What This Book Offers

This isn't your standard math textbook. Instead, it's a friendly guide designed to:

- Introduce you to simple yet powerful mental math techniques
- Build your confidence through gradual, achievable challenges
- Connect math to real-life situations you encounter every day
- Provide strategies to overcome math anxiety
- Make math enjoyable (yes, really!) through games and interesting number facts

How to Use This Book Effectively

To get the most out of this book:

1. Take It at Your Own Pace: There's no rush. Spend as much time as you need on each chapter.
2. Practice Regularly: Even 10 minutes a day can make a significant difference.
3. Apply What You Learn: Try using the techniques in your daily life.
4. Be Kind to Yourself: Remember, mistakes are opportunities to learn, not reasons to give up.
5. Celebrate Small Wins: Every step forward is progress, no matter how small it might seem.

A New Chapter in Your Math Story

As we embark on this journey together, I invite you to open your mind to a new perspective on math. It's not about being fast or perfect; it's about understanding, problem-solving, and seeing the world in new ways.

Remember, every expert was once a beginner. Your past experiences with math don't define your future relationship with it. This book is your opportunity to rewrite your math story, one page at a time.

Are you ready to begin? Let's turn the page and start exploring the fascinating world of mental math!

Chapter 1: Understanding Our Number System

Before we dive into mental math techniques, it's essential to understand the foundation of our number system. This knowledge will make everything else we learn more intuitive and easier to apply.

The Beauty of the Base-10 System

Our number system, known as the decimal or base-10 system, is both simple and powerful. Here's why it's so special:

1. Ten Digits: We use just ten digits (0-9) to represent all numbers, no matter how large.
2. Place Value: Each digit's position in a number determines its value.
3. Powers of 10: Each place value is 10 times greater than the place to its right.

Let's break this down further:

Place Value: The Building Blocks of Numbers

In our base-10 system, each digit in a number represents a multiple of a power of 10. For example, in the number 3,475:

- 5 is in the ones place (5 × 1)
- 7 is in the tens place (7 × 10)
- 4 is in the hundreds place (4 × 100)
- 3 is in the thousands place (3 × 1,000)

So,   3,475 = 3,000 + 400 + 70 + 5

Simple! Right?

Understanding place value is crucial for mental math. It allows us to break down complex calculations into simpler parts.

Practice Set 1.1: Place Value Practice (Solutions at the end of the chapter)

Try breaking down these numbers:

1. 8,392
2. 15,607
3. 240,035

Visualizing Numbers: The Number Line

A number line is a powerful tool for visualizing numbers and their relationships. It's a straight line where each point represents a number, increasing from left to right.

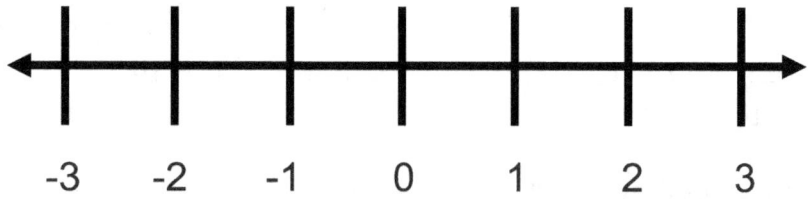

Key points about the number line:

1. It extends infinitely in both directions.
2. The distance between any two consecutive whole numbers is always the same.
3. It includes all numbers: whole numbers, fractions, and irrational numbers.

Visualizing calculations on a number line can make them more intuitive. For example, addition is moving right, subtraction is moving left.

Practice Set 1.2: Number Line Visualization (Solutions at the end of the chapter)

Imagine these calculations on a number line:

1. 5 + 3
2. 7 - 4
3. -2 + 6

Friendly Numbers: Making Calculations Easier

Some numbers are easier to work with than others. We call these "friendly numbers." The most common friendly numbers are:

1. Multiples of 10 (10, 20, 30, 40, ...)
2. Multiples of 5 (5, 15, 25, 35, ...)
3. Multiples of 2 (2, 4, 6, 8, ...)
4. Numbers that make 10 when added together (3 and 7, 6 and 4, ...)

Many mental math techniques involve turning "unfriendly" numbers into friendly ones to make calculations easier.

Practice Set 1.3: Finding Friendly Numbers (Solutions at the end of the chapter)

For each number, find a nearby "friendly number":

1. 48
2. 23
3. 67

The Power of Patterns

Our number system is full of patterns. Recognizing these patterns can make calculations much easier. Here are a few to look out for:

1. Counting by 2s: Even numbers always end in 0, 2, 4, 6, or 8.
2. Counting by 5s: Numbers always end in 0 or 5.
3. Counting by 10s: Numbers always end in 0.

As we progress through this book, we'll discover many more patterns that make mental math easier and more intuitive.

Wrapping Up

Understanding our number system is the first step toward mental math mastery. By grasping place value, visualizing numbers on a number line, recognizing friendly numbers, and spotting patterns, you're building a solid foundation for all the techniques we'll explore in the coming chapters.

Remember, math is all about patterns and relationships between numbers. As you practice, you'll start to see these patterns more easily, making calculations feel more natural and intuitive.

Chapter 1 Practice Sets: Solutions

Practice set 1.1:

1. 8,392 = 8,000 + 300 + 90 + 2
2. 15,607 = 10,000 + 5,000 + 600 + 7
3. 240,035 = 200,000 + 40,000 + 30 + 5

Practice set 1.2:

1. 5 + 3: Start at 5, move 3 steps right to 8
2. 7 - 4: Start at 7, move 4 steps left to 3
3. -2 + 6: Start at -2, move 6 steps right to 4

Practice set 1.3:

1. 48: Nearby friendly numbers include 50 (multiple of 10) or 46 (even number)
2. 23: Nearby friendly numbers include 20 or 25 (multiples of 5)

3. 67: Nearby friendly numbers include 65 (multiple of 5) or 70 (multiple of 10)

In the next chapter, we'll start applying this knowledge to The addition strategies. Get ready to discover how understanding our number system makes adding in your head easier than you ever imagined!

Chapter 2: Addition Strategies

Remember when addition meant stacking numbers vertically and working from right to left? While that method works on paper, mental math offers us more flexible and often faster approaches. In this chapter, we'll explore strategies that make adding numbers in your head easier and even more enjoyable!

The Magic of Making 10s

Remember our friendly numbers from Chapter 1? Ten is perhaps the friendliest number of all. Our brains love working with 10s, so let's use that to our advantage!

Strategy 1: Making 10s

When adding two numbers, try to adjust them to make 10 first.

Example: 8 + 7

Think: "How can I make 10 from 8?" 8 needs 2 more to make 10

Take that 2 from 7 (leaving 5) Now we have 10 + 5 = 15

Let's try another: 9 + 6

Think: "9 needs 1 to make 10" Take 1 from 6 (leaving 5) Now we have 10 + 5 = 15

Practice Set 2.1: Making 10s (Solutions at the end of the chapter)

Try these:

1. 9 + 8
2. 7 + 6
3. 8 + 5

Counting On: The Simple Approach

Sometimes, when one number is small, it's easier to just count on from the larger number.

Strategy 2: Counting On

Start with the larger number and count up by the smaller number.

Example: 43 + 4

Start at 43 Count: 44, 45, 46, 47 The answer is 47

This works best when:

- One number is much larger than the other
- The smaller number is 4 or less

Practice Set 2.2: Counting On (Solutions at the end of the chapter)

Try these:

1. 62 + 3
2. 85 + 2
3. 91 + 4

Breaking Numbers Apart

Sometimes it's easier to break numbers into friendlier parts.

Strategy 3: Breaking Numbers Apart

Split numbers based on place value (tens and ones).

Example: 46 + 38

Split: 46 = 40 + 6

38 = 30 + 8

Add tens: 40 + 30 = 70

Add ones: 6 + 8 = 14

Combine: 70 + 14 = 84

Another example: 25 + 37

Split: 25 = 20 + 5

37 = 30 + 7

Add tens: 20 + 30 = 50

Add ones: 5 + 7 = 12

Combine: 50 + 12 = 62

Practice Set 2.3: Breaking Numbers Apart (Solutions at the end of the chapter)

Try these:

1. 45 + 27
2. 63 + 28
3. 52 + 39

The Compensation Method

Sometimes it's easier to round up to a friendly number and then adjust.

Strategy 4: Compensation

Round one number up to make it friendlier, then subtract the extra.

Example: 58 + 26

Think: "58 is close to 60"

Add 2 to get to 60   60 + 26 = 86

Subtract the extra 2   86 - 2 = 84

Another example: 45 + 38

Think: "38 is close to 40"

Add 2 to get to 40

45 + 40 = 85

Subtract the extra 2   85 - 2 = 83

Practice Set 2.4: Compensation (Solutions at the end of the chapter)

Try these:

1. 68 + 43
2. 76 + 29
3. 94 + 57

Double Plus One

When adding two numbers that are next to each other (like 7 + 8), you can use the doubling strategy.

Strategy 5: Double Plus One

For consecutive numbers, double the smaller number and add 1.

Example: 7 + 8

Double 7: 7 × 2 = 14

Add 1: 14 + 1 = 15

Practice Set 2.5: Double Plus One (Solutions at the end of the chapter)

Try these:

1. 8 + 9
2. 12 + 13
3. 24 + 25

Putting It All Together

The key to becoming good at mental addition is knowing which strategy to use and when. Here's a quick guide:

- For numbers that nearly make 10: Use Making 10s
- When one number is small (1-4): Use Counting On
- For two-digit numbers: Try Breaking Numbers Apart
- When one number is close to a friendly number: Use Compensation
- For consecutive numbers: Use Double Plus One

Remember, there's no "wrong" strategy - use whatever works best for you in each situation!

Solutions to Practice Sets

Practice Set 2.1: Making 10s

1. 9 + 8 = 17 (Take 1 from 8 to make 10, then add 7)
2. 7 + 6 = 13 (Take 3 from 6 to make 10, then add 3)
3. 8 + 5 = 13 (Take 2 from 5 to make 10, then add 3)

Practice Set 2.2: Counting On

1. 62 + 3 = 65
2. 85 + 2 = 87
3. 91 + 4 = 95

Practice Set 2.3: Breaking Numbers Apart

1. 45 + 27 = 72 (40 + 20 = 60, 5 + 7 = 12, 60 + 12 = 72)
2. 63 + 28 = 91 (60 + 20 = 80, 3 + 8 = 11, 80 + 11 = 91)
3. 52 + 39 = 91 (50 + 30 = 80, 2 + 9 = 11, 80 + 11 = 91)

Practice Set 2.4: Compensation

1. 68 + 43 = 111 (70 + 43 = 113, 113 - 2 = 111)
2. 76 + 29 = 105 (76 + 30 = 106, 106 - 1 = 105)
3. 94 + 57 = 151 (94 + 60 = 154, 154 - 3 = 151)

Practice Set 2.5: Double Plus One

1. 8 + 9 = 17 (8 × 2 = 16, 16 + 1 = 17)
2. 12 + 13 = 25 (12 × 2 = 24, 24 + 1 = 25)
3. 24 + 25 = 49 (24 × 2 = 48, 48 + 1 = 49)

Looking Ahead

Now that you've mastered these addition strategies, you'll find that many calculations become easier. In the next chapter, we'll explore subtraction strategies that build on these same principles. Remember, practice makes progress - try using these strategies in your daily life, whether you're adding up grocery items or calculating a tip.

Keep practicing, and don't forget - it's okay to use different strategies for different situations. The goal is to find what works best for you!

Chapter 3: Subtraction Strategies

Subtraction often feels trickier than addition. Many of us learned to "borrow" numbers and cross them out on paper, but that's hard to do in your head! In this chapter, we'll explore mental math strategies that make subtraction easier and more intuitive.

Strategy 1: Counting Up (The Shopkeeper's Method)

Ever notice how cashiers count up from your purchase amount to your payment? This strategy often feels more natural than traditional subtraction.

How It Works:

Instead of subtracting down, count up from the smaller number to the larger number.

Example: 83 - 57

Start at 57 To 60: +3

To 80: +20

To 83: +3

Add the jumps: 3 + 20 + 3 = 26

Think of it like making change:

- From 57 to 60 (add 3)
- From 60 to 80 (add 20)
- From 80 to 83 (add 3)
- Total difference: 26

Practice Set 3.1: Counting Up

Try these:

1. 72 - 48
2. 91 - 73
3. 100 - 67

Strategy 2: Finding Friendly Numbers

Just like in addition, working with friendly numbers makes subtraction easier.

How It Works:

Adjust numbers to make them friendlier, then compensate for the adjustment.

Example: 82 - 39

Think: "39 is close to 40" 82 - 40 = 42 (easier!)

But we subtracted 1 too many So add 1 back: 42 + 1 = 43

Another example: 73 - 48

Think: "48 is close to 50" 73 - 50 = 23

But we subtracted 2 too many So add 2 back: 23 + 2 = 25

Practice Set 3.2: Friendly Numbers

Try these:

1. 64 - 39
2. 92 - 58
3. 45 - 28

Strategy 3: Breaking Apart Numbers

Sometimes it's easier to subtract in chunks using place value.

How It Works:

Break the subtraction into tens and ones.

Example: 74 - 32

Break apart: 70 - 30 = 40

4 - 2 = 2

Combine: 40 + 2 = 42

For trickier ones like 74 - 38:

Break it down: 74 - 30 = 44 (subtract tens first)

44 - 8 = 36 (then subtract ones)

Practice Set 3.3: Breaking Apart

Try these:

1. 85 - 42
2. 67 - 23
3. 93 - 45

Strategy 4: The Distance Method

Think about the distance between numbers on a number line.

How It Works:

Visualize the numbers on a number line and find an easy path between them.

Example: 802 - 796

Instead of subtracting,

think: "From 796 to 800 is 4

From 800 to 802 is 2

Total distance: 4 + 2 = 6"

This is especially useful when numbers are close together!

Practice Set 3.4: Distance Method

Try these:

1. 2,000 - 1,995
2. 503 - 498
3. 1,001 - 994

Strategy 5: Taking Away Too Much and Fixing It

Sometimes it's easier to subtract too much and add back.

How It Works:

Subtract a friendly number that's bigger than needed, then add back the extra.

Example: 84 - 27

First subtract 30: 84 - 30 = 54

But we subtracted 30 not 27

Add back 3: 54 + 3 = 57

Practice Set 3.5: Taking Away and Fixing

Try these:

1. 76 - 38
2. 92 - 44

3. 65 - 37

Real-World Applications

Let's see how these strategies help in everyday situations:

1. Shopping:
   - Item costs $23.45, you have $50
   - Use counting up: $23.45 to $23.50 (+0.05), to $24 (+0.50), to $50 (+26)
   - Change due: $26.55
2. Time Calculations:
   - A meeting ended at 3:45 and started at 2:17
   - Use distance method: 2:17 to 2:20 (+3), to 3:20 (+60), to 3:45 (+25)
   - Duration: 1 hour and 28 minutes
3. Budgeting:
   - Monthly budget $2,000, spent $1,645
   - Use friendly numbers: $2,000 - $1,650 = $350, add back $5
   - Remaining: $355

Solutions to Practice Sets

Practice Set 3.1: Counting Up

1. 72 - 48 = 24 (48→50: +2, 50→70: +20, 70→72: +2)
2. 91 - 73 = 18 (73→80: +7, 80→90: +10, 90→91: +1)
3. 100 - 67 = 33 (67→70: +3, 70→100: +30)

Practice Set 3.2: Friendly Numbers

1. 64 - 39 = 25 (64 - 40 = 24, +1)
2. 92 - 58 = 34 (92 - 60 = 32, +2)
3. 45 - 28 = 17 (45 - 30 = 15, +2)

Practice Set 3.3: Breaking Apart

1. 85 - 42 = 43 (80 - 40 = 40, 5 - 2 = 3)
2. 67 - 23 = 44 (60 - 20 = 40, 7 - 3 = 4)

3. 93 - 45 = 48 (90 - 40 = 50, 3 - 5 = -2)

Practice Set 3.4: Distance Method

1. 2,000 - 1,995 = 5
2. 503 - 498 = 5
3. 1,001 - 994 = 7

Practice Set 3.5: Taking Away and Fixing

1. 76 - 38 = 38 (76 - 40 = 36, +2)
2. 92 - 44 = 48 (92 - 50 = 42, +6)
3. 65 - 37 = 28 (65 - 40 = 25, +3)

Troubleshooting Common Challenges

1. Getting Lost in the Steps?
- Start with smaller numbers until you're comfortable
- Write down intermediate steps if needed
- Practice one strategy until it becomes natural
2. Numbers Too Big?
- Break them into smaller chunks
- Use friendly numbers whenever possible
- Remember, it's okay to use a calculator for very large numbers
3. Not Sure Which Strategy to Use?
- For close numbers: Use the distance method
- For numbers ending in 8 or 9: Use friendly numbers
- For money calculations: Try counting up

Looking Ahead

As you practice these subtraction strategies, you'll find that some work better for you than others. That's perfectly normal! In the next chapter, we'll explore multiplication strategies that will build on these skills.

Remember:

- Practice regularly with real-life numbers
- Start with easier problems and work up
- Use what works best for you
- Celebrate your progress!

Keep practicing, and don't forget - every time you use mental math in real life, you're getting better at it!

Chapter 4: Multiplication Basics

Many people's math anxiety stems from multiplication. Those dreaded times tables! But what if I told you that multiplication doesn't have to be about memorization? In this chapter, we'll explore simple, intuitive ways to multiply numbers that actually make sense.

Understanding Multiplication

Before we dive into techniques, let's understand what multiplication means.

Multiplication as Repeated Addition

Multiplication is just a faster way of adding the same number multiple times.

Example: 4 × 3

Think of it as: 3 + 3 + 3 + 3 Or: Four groups of three

This simple way of thinking about multiplication can make larger problems less intimidating.

The Magic of Multiplying by 10, 100, and 1000

Let's start with the easiest multiplication tricks of all!

Strategy 1: Adding Zeros

When multiplying by 10, 100, or 1000, simply add the corresponding number of zeros.

- To multiply by 10: Add one zero

- To multiply by 100: Add two zeros
- To multiply by 1000: Add three zeros

Examples:

7 × 10 = 70   23 × 100 = 2,300   5 × 1000 = 5,000

Practice Set 4.1: Multiplying by Powers of 10

Try these:

1. 8 × 10
2. 45 × 100
3. 12 × 1000
4. 236 × 10

The Easy-Peasy 5 Times Table

Multiplying by 5 is simpler than you might think!

Strategy 2: Multiply by 10 and Halve

To multiply any number by 5:

1. Multiply it by 10 (add a zero)
2. Divide the result by 2

Example: 26 × 5

26 × 10 = 260

260 ÷ 2 = 130

So, 26 × 5 = 130

Practice Set 4.2: Multiplying by 5

Try these:

1. 14 × 5

2. 32 × 5
3. 48 × 5
4. 86 × 5

Doubling and Halving

This strategy can make many multiplication problems easier.

Strategy 3: Double and Halve

When one number is even, you can double one number and halve the other. The product stays the same!

Example: 25 × 8

Halve 8 to get 4

Double 25 to get 50

Now solve 50 × 4 = 200

Another example: 16 × 15

Double 15 to get 30

Halve 16 to get 8

Now solve 30 × 8 = 240

Practice Set 4.3: Double and Halve

Try these:

1. 12 × 25
2. 18 × 50
3. 24 × 15
4. 32 × 25

Breaking Numbers Apart

Just like with addition, we can break larger numbers into friendlier parts.

Strategy 4: The Distributive Property

Break one number into parts that are easier to multiply.

Example: 23 × 4

Break 23 into 20 + 3 Multiply each part by 4: 20 × 4 = 80 3 × 4 = 12 Add the results: 80 + 12 = 92

Another example: 16 × 7

Break 16 into 10 + 6 10 × 7 = 70 6 × 7 = 42 70 + 42 = 112

Practice Set 4.4: Breaking Numbers Apart

Try these:

1. 34 × 3
2. 42 × 5
3. 26 × 4
4. 83 × 2

The Box Method: A Visual Approach

For those who like to visualize numbers, the box method can be helpful.

Strategy 5: Box Method

Draw a box divided into sections for each place value.

Example: 24 × 13

4. 86 × 5 = 430 (86 × 10 = 860, halve to get 430)

Practice Set 4.3: Double and Halve
1. 12 × 25 = 300 (6 × 50 = 300)
2. 18 × 50 = 900 (9 × 100 = 900)
3. 24 × 15 = 360 (12 × 30 = 360)
4. 32 × 25 = 800 (16 × 50 = 800)

Practice Set 4.4: Breaking Numbers Apart
1. 34 × 3 = 102 (30 × 3 = 90, 4 × 3 = 12, 90 + 12 = 102)
2. 42 × 5 = 210 (40 × 5 = 200, 2 × 5 = 10, 200 + 10 = 210)
3. 26 × 4 = 104 (20 × 4 = 80, 6 × 4 = 24, 80 + 24 = 104)
4. 83 × 2 = 166 (80 × 2 = 160, 3 × 2 = 6, 160 + 6 = 166)

Practice Set 4.5: Box Method
1. 32 × 21 = 672
2. 45 × 12 = 540
3. 23 × 31 = 713

Key Takeaways
1. Multiplication is just repeated addition
2. Breaking numbers into friendly parts makes multiplication easier
3. Visual methods like the box method can help understand the process
4. Practice with real-life situations makes the skills stick

Looking Ahead
In the next chapter, we'll explore more advanced multiplication techniques, including shortcuts for multiplying by 9 or 11, and special numbers like 25. Remember, the goal isn't to memorize everything - it's to understand the patterns and choose strategies that work best for you.

Break it down:

	20	4
10	200	40
3	60	12

Add all parts: 200 + 40 + 60 + 12 = 312

Practice Set 4.5: Box Method

Try these:

1. 32 × 21
2. 45 × 12
3. 23 × 31

Real-Life Applications

Let's practice these strategies in everyday situations:

1. Shopping: Calculating 5 items at $24 each
2. Recipe Scaling: Doubling a recipe that calls for 125ml
3. Party Planning: 25 guests who each need 8 napkins

Solutions to Practice Sets

Practice Set 4.1: Multiplying by Powers of 10

1. 8 × 10 = 80
2. 45 × 100 = 4,500
3. 12 × 1000 = 12,000
4. 236 × 10 = 2,360

Practice Set 4.2: Multiplying by 5

1. 14 × 5 = 70 (14 × 10 = 140, halve to get 70)
2. 32 × 5 = 160 (32 × 10 = 320, halve to get 160)
3. 48 × 5 = 240 (48 × 10 = 480, halve to get 240)

Keep practicing these basic strategies until they feel comfortable. You'll be surprised how quickly they become second nature!

Chapter 5: Advanced Multiplication Techniques

Now that you're comfortable with basic multiplication strategies, let's explore some more advanced techniques that can make larger calculations manageable and even impressive! Don't worry - we'll break everything down into simple, digestible steps.

The Magic of Multiplying by 9

Remember when multiplication by 9 seemed scary? Let's turn it into your favorite party trick!

Strategy 1: The "10 minus 1" Method

Any number multiplied by 9 is the same as that number multiplied by 10, minus the number itself.

- Example: 9×6

    Think: "$10 \times 6 = 60$"

    Subtract one 6: $60 - 6 = 54$

    Therefore, $9 \times 6 = 54$

- Let's try a bigger one: 9×24

    Think: "$10 \times 24 = 240$"

    Subtract one 24: $240 - 24 = 216$

    Therefore, $9 \times 24 = 216$

Practice Set 5.1: Multiplying by 9

Try these:

1. 9 × 7
2. 9 × 13
3. 9 × 31

The Easy 11s

Multiplying by 11 can be surprisingly simple once you know the pattern.

Strategy 2: The "11s Pattern" Method

For two-digit numbers:

1. Add the digits
2. Put the sum between the original digits

Example: 11 × 43

Original number: 43

Add the digits: 4 + 3 = 7

Put 7 between 4 and 3: 473

Therefore, 11 × 43 = 473

Note: If the sum of digits is greater than 9, carry over to the left digit.

Example: 11 × 75

Original number: 75

Add digits: 7 + 5 = 12

Put 2 between, carry 1 to the left 7 + 1 = 8, so: 825

Therefore, 11 × 75 = 825

Practice Set 5.2: Multiplying by 11

Try these:

1. 11 × 24
2. 11 × 52
3. 11 × 68

Squaring Numbers Ending in 5

Here's a neat trick for squaring any number ending in 5.

Strategy 3: The "Square of 5" Method

To square a number ending in 5:

1. Take the tens digit and multiply it by the next whole number
2. Attach 25 to the end

Example: 35^2

Tens digit is 3 Next number is 4

3 × 4 = 12

Attach 25: 1,225

Therefore, $35^2 = 1,225$

Example: 75^2

The tens digit is 7

The next number is 8

7 × 8 = 56

Attach 25: 5,625

Therefore, $75^2 = 5{,}625$

Practice Set 5.3: Squaring Numbers Ending in 5

Try these:

1. 15^2
2. 45^2
3. 85^2

The Box Method for Larger Numbers

When dealing with larger numbers, the box method can make multiplication more manageable.

Strategy 4: The Box Method

Break down each number into place values and multiply parts separately.

Example: 23×46

Break down: $23 = 20 + 3$

$46 = 40 + 6$

Make a box:

	40	6
20	800	120
3	120	18

Add all numbers: $800 + 120 + 120 + 18 = 1{,}058$

Therefore, $23 \times 46 = 1{,}058$

Practice Set 5.4: The Box Method

Try these:

1. 32 × 24
2. 45 × 31
3. 26 × 53

Special Multipliers

Some numbers have special properties that make multiplication easier.

Strategy 5: Special Number Properties

Multiplying by 25

Think of it as multiplying by 100 and dividing by 4.

Example: 25 × 48

48 × 100 = 4,800

4,800 ÷ 4 = 1,200

Therefore, 25 × 48 = 1,200

Multiplying by 15

Think of it as multiplying by 10 and adding half of that.

Example: 15 × 32

32 × 10 = 320

Half of 320 = 160

320 + 160 = 480

Therefore, 15 × 32 = 480

Practice Set 5.5: Special Multipliers

Try these:

1. 25 × 36
2. 15 × 24
3. 25 × 52

Solutions to Practice Sets

Practice Set 5.1: Multiplying by 9

1. 9 × 7 = 63 (70 - 7)
2. 9 × 13 = 117 (130 - 13)
3. 9 × 31 = 279 (310 - 31)

Practice Set 5.2: Multiplying by 11

1. 11 × 24 = 264 (2_4: 2+4=6)
2. 11 × 52 = 572 (5_2: 5+2=7)
3. 11 × 68 = 748 (6_8: 6+8=14, carry 1)

Practice Set 5.3: Squaring Numbers Ending in 5

1. 15^2 = 225 (1×2=2, attach 25)
2. 45^2 = 2,025 (4×5=20, attach 25)
3. 85^2 = 7,225 (8×9=72, attach 25)

Practice Set 5.4: The Box Method

1. 32 × 24 = 768
2. 45 × 31 = 1,395
3. 26 × 53 = 1,378

Practice Set 5.5: Special Multipliers

1. 25 × 36 = 900 (3,600 ÷ 4)
2. 15 × 24 = 360 (240 + 120)
3. 25 × 52 = 1,300 (5,200 ÷ 4)

Real-World Applications

These techniques can be particularly useful in:

- Calculating discounts (especially 15% off)
- Working with quarter portions (25%)
- Estimating larger purchases
- Quick mental checks of calculator results

Looking Ahead

In the next chapter, we'll explore division shortcuts that complement these multiplication techniques. Remember, these advanced strategies might take some time to master - that's completely normal! Practice with smaller numbers first, and gradually work your way up to larger ones.

Quick Tips for Practice:
1. Start with the strategy that feels most natural to you
2. Practice with "friendly" numbers first
3. Use real-world situations to apply these techniques
4. Don't rush - accuracy comes before speed
5. Remember that even math experts use these shortcuts!

Keep practicing, and soon these "advanced" techniques will become second nature!

Chapter 6: Division Shortcuts

Division often gets a bad rap as the most challenging basic math operations. But here's a secret: with the right strategies, division can become manageable and enjoyable! This chapter will explore practical shortcuts that make mental division easier.

Understanding Division

Before we dive into shortcuts, let's refresh our understanding of division:

- The division is sharing into equal groups
- The division is the opposite of multiplication
- Division can be thought of as "How many times does one number go into another?"

Strategy 1: Division as the Inverse of Multiplication

The easiest way to divide in your head is to think about the related multiplication fact.

Example: 56 ÷ 8

Think: "What times 8 equals 56?"

If you know 7 × 8 = 56

Then 56 ÷ 8 = 7

Practice Set 6.1: Using Multiplication Facts

Solve these using related multiplication facts:

1. 72 ÷ 9
2. 42 ÷ 6
3. 63 ÷ 7

Strategy 2: Dividing by 2, 5, and 10

These are your "friendly divisors" - they're easier to work with than other numbers.

Dividing by 2

To divide by 2, just halve the number.

- Even numbers are easy to halve
- For odd numbers, halve the even part and add 0.5

Examples:

86 ÷ 2 = 43 (80 ÷ 2 = 40, 6 ÷ 2 = 3)

75 ÷ 2 = 37.5 (74 ÷ 2 = 37, 1 ÷ 2 = 0.5)

Dividing by 5

To divide by 5, divide by 10 and multiply by 2.

Example: 85 ÷ 5

85 ÷ 10 = 8.5

8.5 × 2 = 17

Dividing by 10

Just move the decimal point one place left.

340 ÷ 10 = 34

567 ÷ 10 = 56.7

Practice Set 6.2: Friendly Divisors

Try these:

1. 94 ÷ 2

2. 165 ÷ 5
3. 820 ÷ 10

Strategy 3: The Chunking Method

For larger divisions, break them down into smaller, manageable chunks.

Example: 156 ÷ 12

Think: "How many 12s make 156?"

Start with what you know: 12 × 10 = 120 (First chunk)

156 - 120 = 36 remaining 12 × 3 = 36 (Second chunk)

So, 10 + 3 = 13

Therefore, 156 ÷ 12 = 13

Another example: 424 ÷ 8

Think in chunks:

8 × 50 = 400 (First big chunk)

424 - 400 = 24 remaining

8 × 3 = 24 (Final chunk)

So, 50 + 3 = 53

Therefore, 424 ÷ 8 = 53

Practice Set 6.3: Chunking Method

Try these:

1. 144 ÷ 12
2. 248 ÷ 8

3. 312 ÷ 6

Strategy 4: Using Known Facts

Sometimes you can use a known division fact to figure out a harder one.

Example: 880 ÷ 11

If you know 88 ÷ 11 = 8

Then 880 ÷ 11 = 80 (Because 880 is 88 × 10)

Practice Set 6.4: Using Known Facts

Try these:

1. 960 ÷ 12
2. 1500 ÷ 15
3. 640 ÷ 8

Strategy 5: Division by 25

Division by 25 has a neat shortcut: multiply by 4 and divide by 100.

Example: 175 ÷ 25

175 × 4 = 700

700 ÷ 100 = 7

So, 175 ÷ 25 = 7

Practice Set 6.5: Division by 25

Try these:

1. 225 ÷ 25
2. 375 ÷ 25
3. 825 ÷ 25

Real-World Applications

Let's look at some practical situations where these division shortcuts come in handy:

1. Splitting Bills
- Dividing a $84 bill between 4 people
- First divide by 2 ($42), then by 2 again ($21 each)
2. Shopping
- Finding the unit price when 5 items cost $15
- 15 ÷ 5 = 3 (divide by 5 shortcut)
3. Cooking
- Halving a recipe that serves 8 people
- Using division by 2 for all ingredients

Solutions to Practice Sets

Practice Set 6.1

1. 72 ÷ 9 = 8 (9 × 8 = 72)
2. 42 ÷ 6 = 7 (6 × 7 = 42)
3. 63 ÷ 7 = 9 (7 × 9 = 63)

Practice Set 6.2

1. 94 ÷ 2 = 47
2. 165 ÷ 5 = 33 (165 ÷ 10 = 16.5, × 2 = 33)
3. 820 ÷ 10 = 82

Practice Set 6.3

1. 144 ÷ 12 = 12 (12 × 10 = 120, 12 × 2 = 24, 10 + 2 = 12)
2. 248 ÷ 8 = 31 (8 × 30 = 240, 8 × 1 = 8, 30 + 1 = 31)
3. 312 ÷ 6 = 52 (6 × 50 = 300, 6 × 2 = 12, 50 + 2 = 52)

Practice Set 6.4

1. 960 ÷ 12 = 80 (96 ÷ 12 = 8, so 960 ÷ 12 = 80)
2. 1500 ÷ 15 = 100 (150 ÷ 15 = 10, so 1500 ÷ 15 = 100)

3. $640 \div 8 = 80$ ($64 \div 8 = 8$, so $640 \div 8 = 80$)

Practice Set 6.5

1. $225 \div 25 = 9$ ($225 \times 4 = 900$, $900 \div 100 = 9$)
2. $375 \div 25 = 15$ ($375 \times 4 = 1500$, $1500 \div 100 = 15$)
3. $825 \div 25 = 33$ ($825 \times 4 = 3300$, $3300 \div 100 = 33$)

Looking Ahead

Now that you've mastered these division shortcuts, you'll find that many calculations become more manageable. In the next chapter, we'll explore percentages and fractions, building on these division skills.

Remember:

- Start with the easiest strategy that fits the problem
- Practice with real-world situations
- It's okay to break bigger numbers into chunks
- Use known facts to solve harder problems

The more you practice these shortcuts, the more natural they'll become. Keep going - you're doing great!

Chapter 7: Percentages and Fractions: Your Everyday Math Tools

If you've ever been shopping during a sale, splitting a restaurant bill, or adjusting a recipe, you've dealt with percentages and fractions. Many people find these concepts intimidating, but by the end of this chapter, you'll see them as helpful tools rather than mathematical obstacles.

Understanding the Connection

First, let's understand how percentages, fractions, and decimals are all connected:

1/2 = 0.5 = 50%

1/4 = 0.25 = 25%

3/4 = 0.75 = 75%

They're just different ways of expressing the same thing!

Quick Percentage Calculations

Strategy 1: Working with 10%

Finding 10% is as easy as moving the decimal point one place left.

Examples:

- 10% of 80 = 8
- 10% of 250 = 25
- 10% of 1,200 = 120

Why this works: 10% is the same as dividing by 10.

Strategy 2: Finding 5%

Half of 10% equals 5%. Once you know 10%, just divide by 2.

Examples:

- 5% of 80 = 4 (half of 8)
- 5% of 250 = 12.50 (half of 25)
- 5% of 1,200 = 60 (half of 120)

Strategy 3: Finding 20%

Double 10% to find 20%.

Examples:

- 20% of 80 = 16 (double 8)
- 20% of 250 = 50 (double 25)
- 20% of 1,200 = 240 (double 120)

Practice Set 7.1: Basic Percentages

Find:

1. 10% of 450
2. 5% of 160
3. 20% of 350

Real-World Application: Calculating Tips

Now let's use these skills for calculating tips:

15% Tip (Nice service)

1. Find 10%
2. Add half of that (5%)

Example: $40 bill

10% = $4

5% = $2

15% = $6 total tip

20% Tip (Great service)
1. Find 10%
2. Double it

Example: $40 bill

10% = $4

Double it: $8 total tip

Practice Set 7.2: Restaurant Tips

Calculate 15% and 20% tips for these bills:

1. $60
2. $85
3. $120

Shopping Discounts

Strategy 4: Finding 25% (Quarter Off)

25% is the same as one-fourth. Find 10% and add half of that amount.

Example: 25% off $80

10% = $8

Half of $8 = $4

10% + 10% + 5% = 25%

$8 + $8 + $4 = $20 discount

Strategy 5: Finding 50% (Half Off)

This one's easy - just divide by 2!

Practice Set 7.3: Shopping Discounts

Find the final price after these discounts:

1. 25% off $120
2. 50% off $85
3. 30% off $200 (Hint: Use 10% × 3)

Working with Simple Fractions

Common Fraction-Decimal-Percentage Conversions

Keep these handy:

1/2 = 0.5 = 50%

1/3 ≈ 0.33 = 33.3%

1/4 = 0.25 = 25%

1/5 = 0.2 = 20%

3/4 = 0.75 = 75%

Strategy 6: Recipe Adjustments

When halving a recipe:

- Divide each number by 2
- 1/2 of 1/4 cup = 1/8 cup
- 1/2 of 1/3 cup ≈ 2.5 tablespoons

When doubling a recipe:

- Multiply each number by 2
- 2 × 1/4 cup = 1/2 cup
- 2 × 1/3 cup = 2/3 cup

Practice Set 7.4: Recipe Adjustments

1. Half a recipe calling for 3/4 cup
2. Double a recipe calling for 1/3 cup
3. Half a recipe calling for 11/2 cups

Mental Math Shortcuts for Percentages

The "Times-Ten-Divide-Ten" Method

To find any percentage:

1. Convert the percentage to a decimal
2. Multiply by 10 to make it easier
3. Divide the result by 10

Example: 35% of 80

35% = 0.35

80 × 10 = 800

800 × 0.35 = 280

280 ÷ 10 = 28

Practice Set 7.5: Mixed Percentage Problems

Calculate:

1. 15% of 240
2. 40% of 150
3. 60% of 80

Solutions to Practice Sets

Practice Set 7.1: Basic Percentages

1. 10% of 450 = 45
2. 5% of 160 = 8

3. 20% of 350 = 70

Practice Set 7.2: Restaurant Tips

1. $60 bill:
   - 15% = $9 (10% = $6, plus half = $3)
   - 20% = $12 (double $6)
2. $85 bill:
   - 15% = $12.75 (10% = $8.50, plus half = $4.25)
   - 20% = $17 (double $8.50)
3. $120 bill:
   - 15% = $18 (10% = $12, plus half = $6)
   - 20% = $24 (double $12)

Practice Set 7.3: Shopping Discounts

1. 25% off $120:
   - Discount = $30 (10% = $12, × 2.5)
   - Final price = $90
2. 50% off $85:
   - Final price = $42.50
3. 30% off $200:
   - Discount = $60 (10% = $20, × 3)
   - Final price = $140

Practice Set 7.4: Recipe Adjustments

1. Half of 3/4 cup = 3/8 cup
2. Double 1/3 cup = 2/3 cup
3. Half of 11/2 cups = 11/4 cup = 2.75 cup(s)

Practice Set 7.5: Mixed Percentage Problems

1. 15% of 240 = 36
2. 40% of 150 = 60
3. 60% of 80 = 48

Real-Life Applications

Remember to use these techniques for:

- Shopping discounts
- Restaurant tips
- Sales tax calculations
- Recipe adjustments
- Budget planning
- Investment returns
- Grade calculations

Looking Ahead

Now that you're comfortable with percentages and fractions, you'll find many daily calculations much easier. In the next chapter, we'll explore estimation techniques that will help you quickly check if your calculations make sense. Remember, practice these skills in real-life situations - that's where they become truly valuable!

Quick Reference Card:

Write this quick reference on a small piece of paper to keep in your wallet:

Common Percentages:

10% = Move decimal left one place

5% = Half of 10%

15% = 10% + 5%

20% = Double 10%

25% = Quarter = 10% + 10% + 5%

50% = Half

Common Fractions:

1/2 = 50%

1/4 = 25%

3/4 = 75%

1/3 ≈ 33.3%   (The ≈ symbol means approximately equal to)

1/5 = 20%

Chapter 8: Estimation and Approximation

Have you ever looked at your shopping cart and wanted to know roughly how much you'll spend before the checkout? Or do you need to quickly figure out if you have enough paint to cover a wall? Welcome to the world of estimation – one of the most practical math skills you'll ever learn!

Why Estimate?

Before we dive into techniques, let's understand why estimation is so valuable:

1. Quick Decision Making: Estimation helps you make rapid decisions without exact calculations
2. Error Checking: It helps you spot when a calculation seems "off"
3. Real-World Practicality: Most real-life situations don't require exact numbers
4. Reduced Math Anxiety: Estimation takes the pressure off getting an "exact" answer

The Art of Rounding

Strategy 1: Rounding to Friendly Numbers

Round numbers to the nearest:

- 10 for numbers under 100
- 100 for numbers between 100 and 1,000
- 1,000 for larger numbers

Examples:

47 → 50

182 → 200

3,678 → 4,000

Practice Set 8.1: Rounding

Round these numbers to the nearest friendly number:

1. 73
2. 891
3. 2,347
4. 16,789

Front-End Estimation

Strategy 2: Focus on the First Digits

For quick estimates, focus only on the first one or two digits.

Example: Estimating 438 + 561

400 + 500 = 900

Real answer: 999

Estimation was close enough for most purposes!

Practice Set 8.2: Front-End Estimation

Estimate these sums:

1. 672 + 428
2. 1,247 + 3,582
3. $23.45 + $18.92

Real-World Applications

Scenario 1: Shopping

Imagine your grocery list:

- Bread $3.99
- Milk $4.75
- Cheese $6.89
- Fruits $12.45
- Vegetables $8.95

Quick estimation:

$4 + $5 + $7 + $12 + $9 = $37

Actual total: $37.03

Scenario 2: Paint Coverage

If paint covers 400 square feet per gallon:

Room dimensions: 15.5 ft × 12.3 ft

Quick estimate: 15 × 12 = 180 square feet

You'll need less than half a gallon

The Clustering Method

Strategy 3: Clustering Similar Numbers

When dealing with multiple numbers, group similar ones together.

Example: Adding up expenses:

$18.45, $19.20, $21.15, $5.60, $4.90

Group similar numbers:

($18 + $19 + $21) ≈ $60

($6 + $5) ≈ $10

Total estimate: $70

Actual: $69.30

Practice Set 8.3: Clustering

Estimate these groups of numbers:

1. $23.45, $24.10, $22.95, $8.45
2. 182, 195, 178, 45, 42
3. 2.8kg, 3.1kg, 2.9kg, 1.2kg

Percentage Estimates

Strategy 4: Quick Percentage Calculations

Use friendly percentages to estimate:

- 10% = divide by 10
- 20% = divide by 5
- 25% = divide by 4
- 50% = divide by 2

Example: 15% tip on $83.50

10% = $8.35

5% = $4.18

15% ≈ $12.50 (rounding $12.53)

Practice Set 8.4: Percentage Estimates

Estimate:

1. 15% of $67.80
2. 18% of $95.00
3. 22% of $125.00

When Close Enough Is Perfect

The "Reasonable Range" Concept

Instead of a single estimate, consider a reasonable range:

Estimating 312 × 5.8

300 × 6 = 1,800

325 × 5.5 = 1,787.50

Reasonable range: 1,780-1,810

Actual answer: 1,809.60

Making Estimation a Habit

Tips for developing estimation skills:

1. Practice estimating before exact calculations
2. Check if your answer seems reasonable
3. Use estimation to catch calculation errors
4. Start with simple situations and build up

Ballpark Figures in Business

Common Business Estimations:

- Project budgets
- Time requirements
- Resource needs
- Market size
- Revenue projections

Example: Project Budget Estimation

Staff costs: ~$5,000/month × 6 months ≈ $30,000

Materials: ~$800/month × 6 months ≈ $5,000

Overhead: ~20% of total ≈ $7,000

Quick estimate: $42,000

Solutions to Practice Sets

Practice Set 8.1: Rounding

1. 73 → 70
2. 891 → 900
3. 2,347 → 2,300
4. 16,789 → 17,000

Practice Set 8.2: Front-End Estimation

1. 672 + 428 ≈ 600 + 400 = 1,000
2. 1,247 + 3,582 ≈ 1,200 + 3,600 = 4,800
3. $23.45 + $18.92 ≈ $23 + $19 = $42

Practice Set 8.3: Clustering

1. ($23 + $24 + $23) + $8 ≈ $70 + $8 = $78
2. (180 + 195 + 180) + (45 + 42) ≈ 555 + 87 = 642
3. (2.8 + 3.1 + 2.9) + 1.2 ≈ 9 + 1.2 = 10.2kg

Practice Set 8.4: Percentage Estimates

1. 15% of $67.80 ≈ $10 (10% = $6.78, 5% = $3.39)
2. 18% of $95.00 ≈ $17 (10% = $9.50, 5% = $4.75, 3% = $2.85)
3. 22% of $125.00 ≈ $28 (20% = $25, 2% = $2.50)

Looking Ahead

Remember, estimation is not about being exactly right – it's about being useful enough for the situation. In the next chapter, we'll explore how to apply these estimation skills specifically to personal finance situations.

Practice your estimation skills daily:

- Estimate your grocery bill while shopping
- Guess the time a task will take
- Estimate distances while driving
- Predict restaurant bills before they arrive

The more you practice estimation, the more natural and accurate your guesses will become. And remember – in many real-world situations, a good estimate is more useful than a precise calculation!

Chapter 9: Mental Math in Daily Life

Now that you've learned various mental math techniques, let's put them to work in real-world situations. This chapter will show you how to apply these skills to make your daily life easier, save money, and even impress your friends!

Personal Finance and Shopping

Quick Percentage Calculations for Shopping

Strategy 1: Calculating Discounts

Most common discounts are 10%, 20%, 25%, 30%, and 50%. Here's how to calculate them quickly:

10% off: Move the decimal point one place left

- $60 → $6 off
- $45 → $4.50 off

20% off: Calculate 10% and double it

- $80 → $8 (10%) → $16 off
- $35 → $3.50 (10%) → $7 off

25% off: Find a quarter (divide by 4)

- $100 → $25 off
- $40 → $10 off

50% off: Divide by 2

- $70 → $35 off
- $25 → $12.50 off

Practice Set 9.1: Discount Calculations

Calculate these discounts:

1. 20% off $65
2. 25% off $80
3. 10% off $95
4. 50% off $130

Sales Tax Quick Calculations

For approximate tax calculations:

- 5% tax: Use the 10% method and divide by 2
- 6% tax: Calculate 5% and add 1%
- 8% tax: Use 10% and subtract 2%
- 9% tax: Use 10% and subtract 1%

Example: $45 item with 8% tax

10% of $45 = $4.50

Subtract 2% ($0.90)

Tax = $3.60

Total = $48.60

Restaurant Tips and Bill Splitting

Quick Tip Calculations

Strategy 2: Standard Tips

- 15% tip: Calculate 10% and add half of that
- 20% tip: Double the 10% amount
- 18% tip: Find 20% and subtract a little

Example: $84 bill

10% = $8.40

20% tip = $16.80

15% tip = $12.60 (8.40 + 4.20)

18% tip = about $15.12

Splitting the Bill

Strategy 3: Division Shortcuts

For splitting bills among 2-6 people:

For 2 people: Divide by 2 For 4 people: Divide by 2 twice For 3 people: Divide by 2, then add half of that For 5 people: Divide by 10, then multiply by 2

Example: Split $120 among 5 people

$120 ÷ 10 = $12

$12 × 2 = $24 per person

Practice Set 9.2: Restaurant Math

1. Calculate 18% tip on $65
2. Split $144 among 4 people
3. Calculate total bill: $85 + 20% tip + 8% tax

Kitchen Calculations

Recipe Adjustments

Strategy 4: Scaling Recipes

Halving a recipe: Divide each quantity by 2 Doubling a recipe: Multiply each quantity by 2 1½ times a recipe: Add half of original to original

Example: Scale this recipe 1½ times:

- 2 cups flour
- 3 tablespoons sugar
- ¾ cup milk

Flour: 2 + 1 = 3 cups

Sugar: 3 + 1.5 = 4.5 tablespoons

Milk: 0.75 + 0.375 = 1.125 cups

Converting Between Units

Common conversions to remember:

- 1 cup = 16 tablespoons
- 1 tablespoon = 3 teaspoons
- 1 cup = 240ml (approximately)
- 1 pound = 16 ounces

Strategy 5: Unit Conversion Shortcuts

To convert between cups and tablespoons:

- 1/4 cup = 4 tablespoons
- 1/3 cup = 5⅓ tablespoons
- 1/2 cup = 8 tablespoons

Practice Set 9.3: Kitchen Calculations

1. Double a recipe that calls for ¾ cup sugar
2. Convert 6 teaspoons to tablespoons
3. Scale this recipe by ½: 2½ cups flour, 4 tablespoons butter

Time Calculations

Adding and Subtracting Time

Strategy 6: Time Math

- Adding hours: Use 12-hour chunks
- Adding minutes: Use 15-minute and 30-minute chunks
- Crossing hour boundaries: Break into smaller parts

Example: What time will it be 2½ hours after 10:45?

10:45 + 2 hours = 12:45

12:45 + 30 minutes = 1:15

Practice Set 9.4: Time Problems

1. A meeting ends at 3:30 PM after 1¾ hours. When did it start?
2. If you leave at 2:15 PM for a 45-minute drive, when will you arrive?
3. How long between 10:45 AM and 2:30 PM?

Household Projects

Area Calculations for Home Improvement

Strategy 7: Quick Area Calculations

For rectangular rooms:

- Round to the nearest foot
- Multiply length × width
- Add back small adjustments

Example: Carpet needed for room 12'8" × 15'4"

Round to 13' × 15'

13 × 15 = 195 square feet

Add small adjustment for exactness

Paint Coverage Calculations

Standard paint covers about 400 sq ft per gallon Quick calculation: Divide total square feet by 400

Practice Set 9.5: Home Projects

1. Calculate the flooring needed for room 16'6" × 12'8"
2. How many gallons of paint for walls 15' × 12' × 8' high (2 coats)?
3. Calculate fencing needed for yard 45' × 30'

Solutions to Practice Sets

Practice Set 9.1: Discount Calculations

1. 20% off $65 = $13
2. 25% off $80 = $20
3. 10% off $95 = $9.50
4. 50% off $130 = $65

Practice Set 9.2: Restaurant Math

1. Calculate 18% tip on $65 = $11.70
2. Split $144 among 4 people = $36 per person
3. Calculate total bill: $85 + 20% tip + 8% tax: $110.16

Practice Set 9.3: Kitchen Calculations

1. Double a recipe that calls for ¾ cup sugar: ¾ cup x 2 = 1 ½ cups sugar
2. Convert 6 teaspoons to tablespoons: 6 teaspoons ÷ 3 (since 1 tablespoon = 3 teaspoons) = 2 tablespoons
3. Scale this recipe by ½: 1 ¼ cups flour, 2 tablespoons butter

Practice Set 9.4: Time Problems

1. A meeting ends at 3:30 PM after 1 ¾ hours. When did it start?
    1 ¾ hours = 1 hour 45 minutes
    3:30 PM - 1 hour 45 minutes = 1:45 PM

2. If you leave at 2:15 PM for a 45-minute drive, when will you arrive?
   2:15 PM + 45 minutes = 3:00 PM

3. How long between 10:45 AM and 2:30 PM?
   2:30 PM - 10:45 AM = 3 hours 45 minutes

Practice Set 9.5: Home Projects

1. Approximately 213 square feet.
2. Need 3 gallons of paint. (Round up for safety margin)
3. Need 150 linear feet of fencing.

Real-World Applications Challenge

Try these techniques in your daily life this week:

1. Calculate shopping discounts without your phone
2. Split a restaurant bill mentally
3. Scale a recipe up or down
4. Estimate arrival time for a journey
5. Calculate a room's area for a home project

Looking Ahead

These practical applications will help cement your mental math skills through regular use. In the next chapter, we'll explore some fun number games and puzzles that can help you further strengthen these abilities while entertaining friends and family!

Remember:

- Start with easier calculations and build up
- Practice in real situations
- Don't worry about being perfect - estimates are often good enough
- Keep your calculator as a backup while building confidence

The more you use these techniques daily, the more natural they'll become. Soon, you'll find yourself reaching for your phone calculator less and less!

Chapter 10: Memory Techniques for Math Facts

Have you ever wondered why some people seem to remember numbers effortlessly while others struggle? The secret isn't a "better math brain" - it's having effective memory techniques. In this chapter, we'll explore powerful strategies to help you remember math facts more easily and reliably.

Understanding How Memory Works

Before diving into specific techniques, let's understand how our memory works with numbers:

1. Short-term memory: Holds information briefly (like a phone number you're about to dial)
2. Working memory: Manipulates information (like when you're calculating a tip)
3. Long-term memory: Stores information for later use (like multiplication tables)

The goal is to move math facts from short-term to long-term memory in a way that makes them easy to recall.

Visualization Techniques

1. Number Pictures

Associate numbers with visual images that represent their shape or value:

- 0 = egg or circle
- 1 = pencil or lamp post
- 2 = swan
- 3 = heart on its side
- 4 = sailboat

- 5 = hook
- 6 = snake
- 7 = boomerang
- 8 = snowman
- 9 = balloon and string

Practice Exercise 10.1: Creating Number Stories

Create a story using these number pictures to remember this sequence: 847 Example: "A snowman (8) threw a boomerang (7) that hit a sailboat (4)"

Try these sequences:

1. 365
2. 942
3. 718

Pattern Recognition

1. Number Patterns in Multiplication

Look for patterns that make multiplication easier to remember:

Pattern: Multiplying by 9

- digits in the answer always add up to 9
- first digit is one less than the number being multiplied

9 × 4 = 36 (3 + 6 = 9)

9 × 5 = 45 (4 + 5 = 9)

9 × 6 = 54 (5 + 4 = 9)

Pattern: Multiplying by 11 (two-digit numbers)

- Add the digits and put the sum between them

11 × 25 = 275 (2 and 5 with their sum 7 in between)

11 × 43 = 473 (4 and 3 with their sum 7 in between)

Practice Exercise 10.2: Pattern Recognition

Find the patterns in these sequences:

1. 12, 24, 36, 48, __
2. 2, 4, 8, 16, __
3. 1, 4, 9, 16, __

Mnemonic Devices

1. Word Association

Create memorable phrases where the first letter of each word represents a number or mathematical concept:

Order of Operations "Please Excuse My Dear Aunt Sally"

- Parentheses
- Exponents
- Multiplication/Division
- Addition/Subtraction

2. Rhymes and Songs

Create or use existing rhymes to remember math facts:

For multiplying by 5:

"Five, ten, fifteen, twenty,

These numbers come so plenty!

Twenty-five and thirty too,

Counting by fives is easy to do!"

Practice Exercise 10.3: Creating Mnemonics

Create your own mnemonics for:

1. The first 6 perfect squares (1, 4, 9, 16, 25, 36)
2. The factors of 24
3. The first 5 prime numbers

Chunking Method

Break larger numbers into smaller, manageable "chunks" that are easier to remember.

Example: Remembering 847593 Break it into:

- 847 (eight forty-seven)
- 593 (five ninety-three)

Or even smaller:

- 84
- 75
- 93

Practice Exercise 10.4: Chunking Numbers

Break these numbers into memorable chunks:

1. 3816427
2. 9405283
3. 6720934

Spaced Repetition System

This is a scientifically proven method for moving information into long-term memory.

The Schedule:

1. First review: Same day
2. Second review: Next day
3. Third review: Three days later
4. Fourth review: One week later

5. Fifth review: Two weeks later
6. Final review: One month later

Creating Your Review System

1. Flashcards:
- Write math facts on index cards
- Sort into boxes labeled "Daily," "Weekly," "Monthly"
- Move cards between boxes based on how well you know them
2. Digital Apps:
- Use spaced repetition apps
- Set reminders for review sessions
- Track your progress

Practice Exercise 10.5: Creating a Review Schedule

Create a 30-day review schedule for:

1. Multiplication tables 6-9
2. Common fractions to decimals
3. Square numbers 1-15

Memory Palace Technique

This ancient technique involves associating information with specific locations in a familiar place.

Steps:

1. Choose a familiar route (like walking through your home)
2. Create vivid images for math facts
3. Place these images at specific points along your route

Example: Remembering Perfect Squares

- Front door: $1^2 = 1$ (single key in lock)
- Living room: $2^2 = 4$ (four-legged table)
- Kitchen: $3^2 = 9$ (nine plates on shelf)
- Bedroom: $4^2 = 16$ (sixteen pillows on bed)

Practice Exercise 10.6: Building Your Math Palace

Create a memory palace for:

1. The first 5 multiples of 7
2. Common fraction-decimal equivalents
3. Important mathematical constants (π, e, etc.)

Putting It All Together

Remember:

1. Use multiple techniques for the same information
2. Practice regularly
3. Start with small chunks of information
4. Build up gradually
5. Review and reinforce

Solutions to Practice Sets

Practice Exercise 10.1: Creating Number Stories

Example solutions using visual associations:

1. 365
- "A triangle (3) rolled like a snake (6) and turned into a hook (5)"
- "A bird (3) ate a cookie (6) on a branch (5)"
2. 942
- "A balloon (9) turned into a swan (4) that became a duck (2)"
- "A snail (9) found a sailboat (4) shaped like a swan (2)"
3. 718
- "A boomerang (7) hit a candle (1) that became a snowman (8)"
- "A cliff (7) had a lighthouse (1) with a snowman (8) nearby"

Practice Exercise 10.2: Pattern Recognition

1. 12, 24, 36, 48, 60
- Pattern: Adding 12 each time (multiplying by 12)
2. 2, 4, 8, 16, 32

- Pattern: Multiplying by 2 each time (powers of 2)
3. 1, 4, 9, 16, 25
- Pattern: Square numbers ($1^2, 2^2, 3^2, 4^2, 5^2$)

Practice Exercise 10.3: Creating Mnemonics

1. Perfect Squares Mnemonic: "One Perfect Ninja Jumps To Success"
- One (1)
- Perfect (4)
- Ninja (9)
- Jumps (16)
- To (25)
- Success (36)
2. Factors of 24 Mnemonic: "Two Tiny Trees Make Big Delicious Pies" (2, 3, 4, 6, 8, 12, 24)
3. First 5 Prime Numbers Mnemonic: "Two Tiny Fives See Seven" (2, 3, 5, 7, 11)

Practice Exercise 10.4: Chunking Numbers

1. 3816427
- 381-64-27
- "Three eighty-one, sixty-four, twenty-seven"
- Or: 38-16-427 (sports jersey numbers)
2. 9405283
- 940-52-83
- "Nine forty, fifty-two, eighty-three"
- Or: 94-052-83 (year-code-age)
3. 6720934
- 672-09-34
- "Six seventy-two, zero nine, thirty-four"
- Or: 67-209-34 (age-room-number)

Practice Exercise 10.5: Review Schedule Example

30-Day Review Plan:

- Days 1-5: Focus on one table/concept per day
- Days 6-10: Quick review of two items per day
- Days 11-20: Practice with mixed problems
- Days 21-30: Speed challenges and real-world applications

Example for Multiplication Tables 6-9:

- Week 1: Learn one table per day, practice with games
- Week 2: Mix two tables daily
- Week 3: Real-world problems using these tables
- Week 4: Speed drills and mental math challenges

Practice Exercise 10.6: Math Palace Locations

1. Multiples of 7 Memory Palace:
- Front Door: 7 (welcome sign)
- Living Room: 14 (TV channels)
- Kitchen: 21 (drinking age)
- Bedroom: 28 (lunar cycle)
- Bathroom: 35 (shower temperature)

2. Common Fraction-Decimal Locations:
- Kitchen Sink: ½ = 0.5 (half-full)
- Refrigerator: ¼ = 0.25 (quarter gallon)
- Microwave: ¾ = 0.75 (cooking time)
- Oven: ⅓ = 0.33 (temperature dial)
- Table: ⅕ = 0.2 (place settings)

3. Mathematical Constants Palace:
- Study Desk: π = 3.14 (circular coffee stain)
- Bookshelf: e = 2.718 (book arrangement)
- Wall Clock: $\sqrt{2}$ = 1.414 (time showing)
- Window: φ = 1.618 (golden ratio frame)

Remember: These solutions are examples. Personal associations often work better, so encourage creating unique memory devices that resonate with individual experiences and interests.

Looking Ahead

In the next chapter, we'll explore techniques for overcoming math anxiety. The memory techniques you've learned here will help build your confidence as you tackle more complex calculations.

Remember, everyone's memory works differently. Experiment with these techniques and adapt them to what works best for you. The key is finding methods that make sense to your brain and using them consistently.

Keep practicing, and don't forget to celebrate your progress. Every math fact you remember is one step closer to math confidence!

Chapter 11: Overcoming Math Anxiety

By now, you've learned various mental math techniques and strategies. But there's one more crucial skill to develop: managing math anxiety. In this chapter, we'll explore practical ways to build confidence and overcome math-related fears.

Understanding Math Anxiety

What Is Math Anxiety?

Math anxiety is more than just disliking math. It's a real emotional response that can manifest as:

- Racing heart
- Sweaty palms
- Blank mind
- Feeling of panic
- Avoidance of math-related situations
- Negative self-talk about math abilities

Remember: Having math anxiety doesn't mean you're bad at math. Many successful people, including some mathematicians, have dealt with math anxiety.

The Science Behind Math Anxiety

Understanding what happens in your brain during math anxiety can help you manage it better:

1. Fight or Flight Response
   - Your amygdala (emotional center) activates
   - Stress hormones release
   - Working memory becomes limited
   - Logical thinking becomes harder

2. The Cycle of Anxiety

Anxiety → Poor Performance → More Anxiety → More Poor Performance

Breaking this cycle is key to improving your math skills.

Recognizing Your Triggers

Everyone's math anxiety triggers are different. Take a moment to identify yours:

Common Triggers Checklist

Being put on the spot
Time pressure
Complex calculations
Fear of making mistakes
Negative past experiences
Comparison with others
Specific types of math problems

Exercise 11.1: Identifying Your Triggers

Write down:

1. Three situations that make you anxious about math
2. Physical symptoms you experience
3. Thoughts that come to mind

Practical Strategies for Managing Math Anxiety

1. Breathing Techniques

The "4-7-8" Breathing Method:

1. Inhale for 4 counts
2. Hold for 7 counts

3. Exhale for 8 counts
4. Repeat 3-4 times

Use this before attempting math problems or when feeling anxious.

2. Positive Self-Talk

Replace negative thoughts with positive ones:

Instead of	Try Thinking
"I'm bad at math"	"I'm learning and improving"
"I'll never get this"	"I can take my time to understand"
"This is too hard"	"This is challenging, but I can break it down"
"I always make mistakes"	"Mistakes help me learn"

Exercise 11.2: Reframing Negative Thoughts

Write down three negative math-related thoughts you have, then reframe them positively.

3. Progressive Desensitization

Start with math activities that cause minimal anxiety and gradually work up to more challenging ones:

Level 1: Simple counting or addition Level 2: Basic mental math in private Level 3: Using math with family/friends Level 4: Using math in public Level 5: Teaching math concepts to others

4. Growth Mindset Exercises

1. Journal Your Progress
   - Write down small wins
   - Document improvements
   - Reflect on learning moments
2. Celebrate Mistakes
   - Keep a "Learning from Mistakes" log
   - Analyze what went wrong
   - Plan how to approach similar problems

Exercise 11.3: Progress Journal

Start a weekly math journal noting:

- What you learned
- Challenges you faced
- How you overcame them
- Your feelings about math that week

Setting Realistic Goals

SMART Goals for Math Practice

- Specific: "I will practice mental addition for 10 minutes daily"
- Measurable: "I will complete 5 practice problems each day"
- Achievable: Start with easier problems and progress gradually
- Relevant: Choose goals that matter to your daily life
- Time-bound: Set specific timeframes for your goals

Exercise 11.4: Creating Your Math Goals

Write three SMART goals for your math practice.

Emergency Toolkit for Math Anxiety Moments

Keep these strategies handy for when anxiety strikes:

1. The 5-4-3-2-1 Grounding Technique
- Name 5 things you can see
- 4 things you can touch
- 3 things you can hear
- 2 things you can smell
- 1 thing you can taste
2. Quick Calming Strategies
- Take three deep breaths
- Count backward from 20
- Stretch or shake out the tension
- Drink water
- Use positive self-talk
3. Problem-Solving Steps
- Read the problem twice
- Break it into smaller parts
- Use estimation first
- Check your work
- Ask for help if needed

Building a Support System

1. Find a Math Buddy
- Someone to practice with
- Share struggles and successes
- Accountability partner
2. Create a Safe Learning Environment
- Quiet, comfortable space

- No time pressure
  - Resources readily available
  - Permission to make mistakes
3. Use Available Resources
  - Math apps
  - Online tutorials
  - Study groups
  - Professional help if needed

Remember:

1. Recovery from math anxiety is a journey, not a destination
2. Progress isn't always linear
3. Everyone learns differently
4. It's okay to take breaks
5. Seeking help is a sign of strength, not weakness

Looking Forward

As you continue your math journey, remember that anxiety is normal but manageable. Use the strategies in this chapter alongside the mental math techniques you've learned. With practice and patience, you can build both your math skills and your confidence.

Daily Affirmations for Math Confidence

- I can learn and improve at math
- Mistakes are opportunities to learn
- I take math one step at a time
- My brain is capable of learning math
- I celebrate my progress, no matter how small

Remember: You're not alone in this journey. Many people have overcome math anxiety to develop strong math skills. You can too!

Chapter Review Questions

1. What are your main math anxiety triggers?
2. Which breathing technique works best for you?
3. What positive self-talk statements resonate with you?
4. How will you track your progress?
5. What support system can you build?

Keep this chapter handy and refer back to it whenever you need encouragement or strategies for managing math anxiety. You're doing great!

Chapter 12: Fun with Numbers

Introduction

Congratulations on making it to the final chapter! By now, you've learned numerous mental math techniques and built your confidence with numbers. This chapter is your reward - it's time to discover how math can be genuinely fun and even impressive. We'll explore mathematical games, puzzles, and some amazing "tricks" that will not only entertain but also reinforce everything you've learned.

Part 1: Mathematical Party Tricks

The Amazing Number 9

Let's start with some fascinating properties of the number 9 that will amaze your friends and family.

Trick 1: The Multiply by 9 Finger Method

This trick works for multiplying single digits by 9:

1. Hold up both hands with all ten fingers extended
2. For 9 × N, count N fingers from the left and bend that finger down
3. The number of fingers before the bent finger is the tens digit
4. The number of fingers after the bent finger is the ones digit

Example for 9 × 4:

- Count 4 fingers from left and bend the 4th finger
- 3 fingers before = 3 (tens digit)
- 6 fingers after = 6 (ones digit)
- Result: 36

Trick 2: The Magic Sum

Ask someone to:

1. Think of any number
2. Multiply it by 9
3. Add all digits of the result together

The sum of the digits will always be 9 or a multiple of 9!

Example:

- Choose 456
- 456 × 9 = 4,104
- 4 + 1 + 0 + 4 = 9

Part 2: Number Games

Game 1: Target Number

Players: 2 or more Materials: Paper and pencil (or just mental math!)

Rules:

1. Choose a target number between 100 and 999
2. Players take turns adding any number from 1-10
3. First player to reach the exact target number wins
4. Going over the target means you lose your turn

Strategy tip: Work backwards using subtraction to plan your moves!

Game 2: Factor Finding Race

Players: 2 or more Materials: Timer, paper (optional)

Rules:

1. Choose a number between 50-100
2. Players race to find all factors of the number

3. First player with a complete, correct list wins

Example number: 60 Factors: 1, 2, 3, 4, 5, 6, 10, 12, 15, 20, 30, 60

Game 3: The 24 Game

Players: Any number Materials: Playing cards (Ace-9 only)

Rules:

1. Deal 4 cards face up
2. Using each number exactly once and any operations (+, -, ×, ÷), try to make 24
3. First person to find a solution wins that round

Example: Cards shown: 4, 7, 8, 8 Solution: (8 × 4) - (8 ÷ 7) = 24

Part 3: Number Patterns and Puzzles

The Fibonacci Sequence

Each number is the sum of the two before it: 1, 1, 2, 3, 5, 8, 13, 21, 34...

Challenge: How quickly can you calculate the next 3 numbers in the sequence?

Number Pyramids

```
        ?
      7   8
    3   4   5
  1   2   3   4
```

Fill in the top number by adding adjacent pairs moving upward.

Magic Squares

Create a 3×3 grid where all rows, columns, and diagonals sum to the same number. Example:

$$2\ 7\ 6$$

$$9\ 5\ 1$$

$$4\ 3\ 8$$

(All lines sum to 15)

Part 4: Real-World Math Challenges

The Shopping Game

Next time you're at a store:

1. Round prices to the nearest dollar
2. Keep a running total as you shop
3. Try to get within $5 of your actual total

The Time Zone Challenge

Practice calculating time differences:

- If it's 3:15 PM in New York, what time is it in:
  - Los Angeles (- 3 hours)
  - London (+ 5 hours)
  - Tokyo (+ 13 hours)

The Recipe Scaling Game

Take any recipe and:

1. Double it
2. Cut it in half
3. Make it for 3 people (if it serves 4)

Part 5: Digital Math Games and Apps

While this book focuses on mental math, here are some recommended digital resources for further practice:

1. Number-based puzzle games:
   - Sudoku
   - 2048
   - Threes!
2. Math-focused apps:
   - Brilliant
   - Elevate
   - Math Games

Conclusion: Making Math a Daily Adventure

Remember, math isn't just about calculations - it's about seeing patterns, solving puzzles, and having fun with numbers. Try to incorporate these games and challenges into your daily life:

1. Play a number game during your commute
2. Challenge family members to mental math contests
3. Create your own mathematical puzzles
4. Share these tricks with friends and family

The more you play with numbers, the more comfortable and confident you'll become. Math anxiety transforms into math enthusiasm when we approach numbers with a spirit of play and discovery.

Chapter Challenges

1. Create your own number game using the mental math techniques from this book
2. Teach someone else one of the number tricks
3. Try to complete a game of Target Number using only mental math

4. Create your own magic square
5. Time yourself solving number pyramids and track your improvement

Remember: The goal isn't just to get the right answer - it's to have fun while doing it!

Conclusion: Your Ongoing Math Journey

Reflecting on Your Progress

Take a moment to think back to when you first opened this book. Remember those initial feelings about math? Perhaps there was anxiety, uncertainty, or even fear. Now, after working through these chapters, pause and celebrate how far you've come. Whether you've mastered every technique or are still working on building confidence, you've already proven something incredibly important: you can learn and grow in math at any age.

Key Takeaways from Our Journey

Let's recap some of the most important insights we've discovered together:

1. The Power of Mindset

- Math ability isn't fixed; it grows with practice and understanding
- Mistakes are valuable learning opportunities
- Everyone can develop mental math skills, regardless of past experiences

2. Essential Strategies We've Learned

- Making friends with numbers by finding patterns and relationships
- Breaking down complex calculations into manageable parts
- Using estimation to check our work and make quick calculations
- Applying mental math shortcuts in real-world situations

3. Real-World Applications

- Quick calculations while shopping
- Confident budget management
- Effortless bill splitting at restaurants
- Recipe adjustments in cooking
- Time and distance calculations
- Investment and financial decisions

Building Lasting Confidence

Remember, true confidence in math isn't about knowing everything or being the fastest calculator in the room. It's about:

1. Trusting Your Process: Having reliable methods you can fall back on
2. Being Comfortable with Approximation: Knowing when precision matters and when estimates are sufficient
3. Learning from Mistakes: Using errors as stepping stones to deeper understanding
4. Asking Questions: Maintaining curiosity about numbers and patterns
5. Applying Knowledge: Regularly using math skills in daily life

Your Mathematical Toolkit

Through this book, you've developed a versatile set of tools:

- Mental Math Strategies: Various techniques for different types of calculations
- Problem-Solving Approaches: Methods for breaking down complex problems
- Estimation Skills: Quick ways to approximate answers
- Confidence-Building Practices: Techniques for managing math anxiety
- Real-World Applications: Practical ways to use math in daily life

Continuing Your Growth

Your math journey doesn't end with the last page of this book. Here are ways to continue developing your skills:

Daily Practice Ideas

1. Morning Math Ritual: Start your day with a few mental calculations
2. Shopping Challenge: Calculate totals and discounts while shopping
3. Number Games: Play Sudoku, KenKen, or other number puzzles
4. Teaching Others: Share your knowledge with family or friends
5. Math Journal: Keep track of interesting patterns you notice

Resources for Further Learning

- Online math communities and forums
- Math puzzle books and apps
- Educational websites and videos
- Local adult education classes
- Math-related podcasts and blogs

Setting Future Goals

Consider setting some personal math goals:

Short-term Goals (Next 30 Days)

- Practice one new technique each week
- Solve daily mental math problems
- Use mental math in at least one real situation daily

Medium-term Goals (3-6 Months)

- Master all basic calculations without a calculator
- Help others with their math challenges
- Take on more complex mental math challenges

Long-term Goals (1 Year and Beyond)

- Maintain confidence in all basic math situations
- Continue learning new mathematical concepts
- Perhaps mentor others on their math journey

A Personal Note to You

As we conclude this book, I want to share something important: your math journey is uniquely yours. Don't compare your progress to others. Each step forward, no matter how small it might seem, is significant. You've already proven that you can learn and grow – that's an incredible achievement.

Remember, math is not just about numbers; it's about seeing patterns, solving puzzles, and understanding the world around us in new ways. Every time you use these skills, you're not just doing math – you're exercising your brain, building confidence, and opening new doors of opportunity.

Your Next Steps

1. Review: Regularly revisit chapters that challenge you
2. Practice: Keep using these techniques in daily life
3. Connect: Share your progress with others
4. Explore: Look for new mathematical challenges
5. Celebrate: Acknowledge your growth and achievements

Final Thoughts

You've completed this book, but in many ways, your real journey is just beginning. Mathematics is a vast and beautiful subject, and you now have the tools to explore it with confidence. There will still be challenges ahead, but you've proven that you can overcome them.

Remember: You are capable of learning math. You can solve problems creatively. You have the power to keep growing.

Thank you for letting me be part of your mathematical journey. Now go forth and calculate with confidence!

Quick Reference Guide:

Keep these mantras handy for when you need encouragement:

> "Every mistake is a learning opportunity"
>
> "I can break any problem into manageable parts"
>
> "It's okay to estimate and approximate"
>
> "I grow stronger with each calculation"
>
> "Math is a tool I can use to improve my life"

Remember, you're not just learning math – you're transforming your relationship with numbers, one calculation at a time. Keep exploring, keep growing, and most importantly, keep believing in yourself!

Glossary of Terms

Addition Strategies

Methods that simplify adding numbers in your head, such as "Making 10s," "Counting On," and "Breaking Numbers Apart."

Base-10 System

Our number system, where place values increase by powers of 10 (e.g., ones, tens, hundreds).

Box Method

A visual approach to multiplication that breaks numbers into place values, multiplies each part, and then adds them together.

Clustering

An estimation technique where similar numbers are grouped and calculated together for faster results.

Compensation

A mental math strategy where you adjust one number to make the calculation easier, then compensate for that adjustment.
Example: To calculate 58 + 26, adjust 58 to 60, then subtract the extra 2 later: 60+26 −2=84.

Distributive Property

A property of multiplication that allows breaking one number into parts, multiplying each part, and summing the results.
Example: 23 × 4 = (20 × 4) + (3 × 4) = 80 × 12 = 92

Doubling and Halving

A multiplication shortcut where you double one number and halve the other to make calculations easier.
Example: 16×25=8×50=400.

Estimate

A rough calculation of a number or result, used to save time or check accuracy. Often achieved by rounding or simplifying numbers.

Friendly Numbers

Numbers that are easy to work with, such as multiples of 10, 5, or 2. Adjusting calculations to include friendly numbers simplifies math.

Mental Math

Performing calculations in your head without writing them down or using tools like a calculator.

Number Line

A visual representation of numbers on a straight line, used to help with addition, subtraction, and understanding number relationships.

Place Value

The value of a digit depending on its position in a number.
Example: In 3,475, the 3 is in the thousands place, so its value is 3,000.

Practice Sets

Exercises provided at the end of sections to help reinforce the techniques and concepts covered.

Rounding

Adjusting a number to the nearest friendly number (e.g., nearest 10, 100, or whole number) to make calculations easier.

Square of Numbers Ending in 5

A shortcut for squaring numbers ending in 5. Multiply the tens digit by the next whole number, then attach 25.
Example: $35^2 = (3 \times 4)25 = 1225$

Subtraction Strategies

Techniques like "Counting Up" and "Breaking Apart Numbers" to simplify subtraction in mental math.

Visualization

Using tools like number lines, diagrams, or mental images to understand and solve math problems more intuitively.

www.ingramcontent.com/pod-product-compliance
Lightning Source LLC
Chambersburg PA
CBHW071055240526
45469CB00006BD/2305